Amazing World of Night Creatures

Written by Janet Craig

Illustrated by Jean Helmer

Troll Associates

Library of Congress Cataloging-in-Publication Data

Amazing world of night creatures.

Summary: Introduces the characteristics and behavior of such creatures of the night as the kiwi, bat, and raccoon.
1. Nocturnal animals—Juvenile literature.
[1. Nocturnal animals] I. Helmer, Jean Cassels, ill.
II. Title.
QL755.5.P35   1990   591.5   89-5002
ISBN 0-8167-1749-4 (lib. bdg.)
ISBN 0-8167-1750-8 (pbk.)

Copyright © 1990 by Troll Associates, Mahwah, New Jersey
All rights reserved. No part of this book may be used or reproduced in any manner whatsoever without written permission from the publisher.
Printed in the United States of America.

10 9 8 7 6 5 4 3 2 1

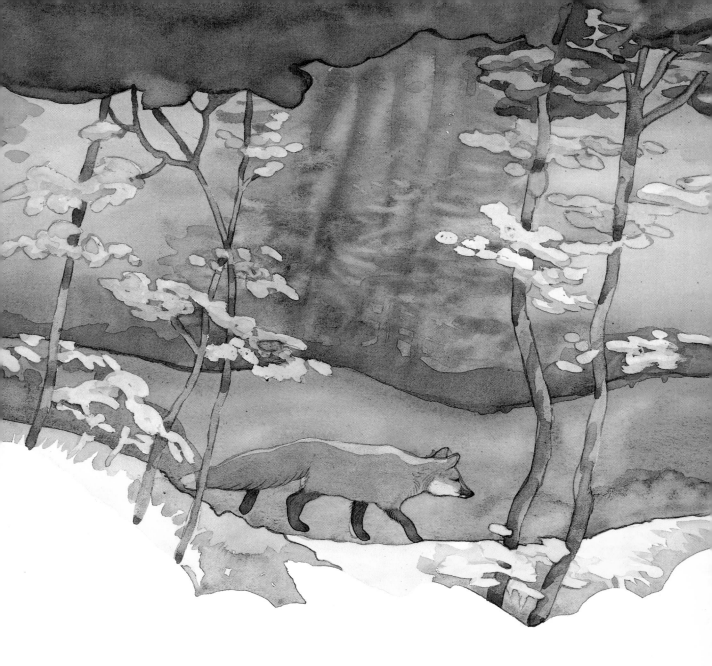

Some fly swiftly through the night. Others prowl on padded feet.

Who are these mysterious creatures? They are the animals of the night. And while you are sound asleep, they are awake!

When night draws near, most of the daytime animals you know go to sleep. But another group of creatures, hidden within their homes, is just awakening. For these animals, nighttime is a time of activity. It is a time for finding food or for escaping danger.

To be active at night, these creatures need very sharp senses. In fact, the senses of most night animals are far sharper than those of many day creatures. Day animals, which are called *diurnal* creatures, can see more easily in the light to hunt or find food. Night animals, called *nocturnal* creatures, need keen senses to live in a world of darkness.

To do this, nature has helped them in many ways. Owls and cats have excellent eyesight to help them hunt in the dark. Rabbits and gerbils have good ears to help them hear when danger is near.

Raccoons are very well suited to the evening. Their brownish-black fur blends into the night so they cannot be seen by their enemies. A raccoon's nose is a wonderful detective—it leads the raccoon to nearby food. And a raccoon's sense of touch is also very good. Its paws are so quick that a raccoon can dip them into the water to catch a fish.

Why are some animals nocturnal? If daylight makes it easier to see and find food, why aren't all creatures awake during the day?

There are several reasons why the night is a better time for certain animals to be active. For some, the heat of the day is too harsh—it would dry out such animals. Snails and slugs, for example, have soft and slimy bodies, which need to stay cool and wet. All day, these animals rest beneath rocks or in safe, shady places. When darkness comes, the air becomes cooler and moister. This is just the right condition for slugs and snails.

Slowly, they creep and crawl from their hiding places to look for food. The snail pokes its head from the safety of its shell. The slug, which looks like a snail without a shell, creeps along, leaving a glistening, slimy trail wherever it goes.

Some creatures like the safety of darkness. Certain rabbits and mice look for seeds and berries to eat at night. They open their large, keen ears to listen for danger. The darkness makes it easier for these animals to escape a hungry fox or weasel.

Another reason many animals are nocturnal is that there is more food for them to find at night. This is especially true of night *predators*, the creatures that hunt and eat other animals. An owl is a night predator. It hunts many of the same animals, such as mice and other rodents, that hawks and eagles catch in the daytime. If owls, hawks, and eagles all hunted during the day, they'd be fighting over the same food.

Most owls, with their excellent eyesight, can hunt in the dark, while the daytime hunters are sleeping. In this way, nature allows two groups of animals—the night and day creatures—to make use of the earth and the food it has to offer.

The owl has special ways of finding and catching its prey. An owl's eyes can see about ten times better than yours can at night.

Why does an owl see better when your eyes and an owl's eyes are alike in many ways? Both have a *lens* in front. The lens collects light and focuses the light rays at the back of the eye. Both people and owls have special cells called *rods* at the back of their eyes. The rods help the eye to see in dim light. But an owl's eye has many more rods than a human eye does. That's how the owl's eye is able to gather more light at night and see better. The night seems very dark to you; but an owl sees the same night in a much brighter fashion.

The owl also has very soft feathers. Because of this, its wings make little or no noise when flying. In this way, the bird is able to surprise its prey.

Owls fly close to the ground when they hunt. Their sharp sense of hearing tells them where a mouse is hiding. Down swoops the owl, grabbing the unlucky mouse in its sharp claws.

Bats are one of the best-known creatures of the dark. They are the only mammals able to fly—and they are the best night fliers of all!

By day, bats sleep hanging upside down in dark caves or other hidden places. When night falls, bats take flight. Thousands have been seen together as they fly from their cave to go hunting.

Imagine how difficult it is to fly in the dark without bumping into things. That is why there are far fewer nighttime fliers than there are daytime fliers. Yet most bats can fly about quickly and change direction in the dark, never once hitting anything. How do they do it? They use something called *echolocation*.

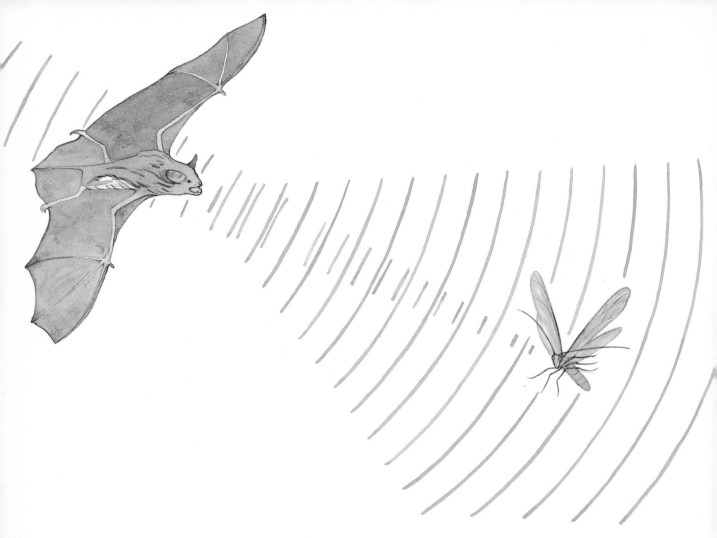

Here is how echolocation works: As a bat flies, it makes very high squeaking sounds. These sounds are so high that people cannot even hear them.

As these squeaking sounds travel through the air, they hit things in their way. When this happens, the sounds bounce off what they have hit, echoing back to the bat. The bat is able to tell from the echoing sounds exactly where the object is.

In this way, the bat can keep from flying into branches or other things in its way. Echolocation also helps the bat find food by letting it know when moths or other insects are nearby.

One special bat, called the fishing bat, uses echolocation to find out if a fish is swimming near the water's surface. Ready for a tasty meal, the bat swiftly dips down and catches the fish in its claws.

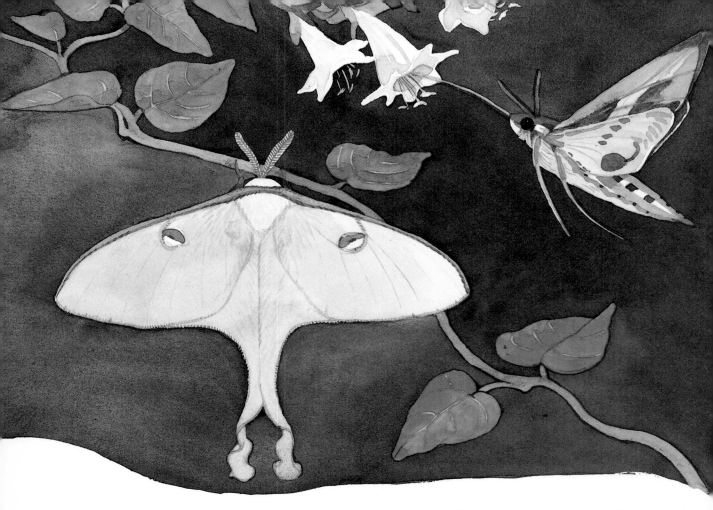

　Creatures both large and small are night animals. Some of the smallest are insects. Most moths are active at night. These flying insects look like butterflies, but most butterflies are active only during the day.
　The pretty white hawk moth sips nectar from flowers that bloom at night. This moth has a very long tongue, which it uses like a straw to find the sweet liquid.
　Some moths give off powerful scents in the night air to find one another. A male luna moth has special feelers on its head. These feelers can smell the scent given off by a female luna moth miles away.

Have you ever looked outside on a summer evening and seen the tiny, twinkling lights of fireflies? The bodies of these insects have special chemicals that make the flashing lights. Fireflies find their mates by blinking these lights.

It's fun to catch a firefly and watch its small light. But after a while, let your firefly go free, for it cannot live very long in a container.

Beavers are busy night animals. They build and fix their dams in the dark. If the dam has a leak, the beaver listens carefully to the sound of water rushing through the hole. Then the beaver drags logs and twigs into place to fix the leak.

People once thought that hyenas—animals that look like dogs—were scavengers. This means that the hyenas would not hunt their own meat, but would only eat meat that was killed and left behind by another animal.

Scientists studied the hyena more closely. They were surprised to find that at night, some hyenas gather together in groups called *packs*. A pack of about twenty-five hyenas may hunt together. They often chase and capture baby zebras or other animals.

Many kinds of cats are also excellent nighttime hunters. Leopards move silently through the woods to surprise their prey, for they have big, soft pads on their paws.

Have you ever seen the way a cat's eyes seem to glow at night? This is because there is a sort of "mirror" at the back of the eye. It is called a *tapetum*. When a ray of light bounces off the tapetum, the rods in the cat's eye see the ray of light twice. The light bouncing off the tapetum creates the "glowing" look.

Being able to see each ray of light twice means the cat sees twice as much light in the dark as we do. These special eyes make the cat one of the best hunters of the night.

Which birds are nocturnal? There are many, but one of the most interesting is the kiwi of New Zealand. This bird, whose wings are too small to enable it to fly, lives in a hole called a *burrow* during the day. At night, it comes out and searches for earthworms to eat. With its head and long, pointed beak bent close to the ground, the kiwi sniffs loudly. Its super sense of smell tells it where the juicy worms are hiding.

Many birds migrate at night. A *migration* is a trip the birds make every year, usually to go from cold weather to warmer lands. By flying at night, the birds can escape hawks that like to feed on them.

The night heron is an excellent fisher. Surrounded by darkness, it stands completely still in the water, waiting for a fish to come near. When one does, the bird quickly grabs it in its sharp beak.

Another creature that glides through the night sky is the flying squirrel. This small animal, which is almost never seen in the daylight, does not really fly. It opens the folds of skin along the sides of its body and leaps into the air. The folds fill with air and help the squirrel glide from tree to tree.

Certain snakes are nighttime hunters. They have special holes on the roofs of their mouths that are sensitive to smell. When a mouse scurries by, it leaves a trace of itself in the air. The snake's sense of smell helps it to track the mouse.

Even in the dark waters of lakes and oceans, nighttime animals are busy. The catfish is one such hunter. It is named for the long feelers on its face, which look like a cat's whiskers. These feelers are not really whiskers. They are called *barbels*, and the catfish uses them to find its way along the bottom of a river as it looks for food.

Often people say the night is calm and quiet. But is this really so? Listen to the sounds of the night—you may be surprised at how noisy they are.

Ribbit! A male frog croaks loudly to its mate. A special air pouch on the frog's throat helps it make these deep sounds. The buzzing of mosquitoes and the chirping of crickets fills the air. The lonely howl of a coyote is heard. And a night bird may add its own lovely song.

Tonight as you go to sleep, think of the special world that is just coming alive. It is a world filled with fascinating animals, and each one is uniquely suited to a life of darkness. For as the daytime creatures sleep, the creatures of the night are busy!